SHARKS

A MIGHTY BITE-Y HISTORY

SHARKS

A MIGHTY BITE-Y HISTORY

words by
MIRIAM FORSTER

pictures by
GORDY WRIGHT

ABRAMS BOOKS FOR YOUNG READERS

NEW YORK

INTRODUCTION

Sharks have swum our oceans for more than 450 million years. That's a long, long, long time, as you can see by these two very scientific lists.

THINGS THAT ARE OLDER THAN SHARKS: A RANDOM LIST

1. BUGS **2. MOSS** **3. STARFISH** **4. JELLYFISH**

5. SEA SPONGES **6. FUNGUS** **7. WORMS**

THINGS THAT ARE YOUNGER THAN SHARKS: A RANDOM LIST

1. HUMANS **2. DOGS** **3. WHALES**

4. DINOSAURS **5. FLOWERS** **6. CROCODILES** **7. TREES**

Sharks are some of the oldest animals on earth today. They've survived extinction and predators and changing climates, and they're so well adapted to their environment that there are are more than five hundred different species of sharks today. But what makes sharks so special?

Well, it turns out, sharks have a very special toolbox. Not the kind with hammers and saws, although some sharks have those too. The shark toolbox has special teeth, special senses, and even special skin!

In this book, we'll take a look at sharks, both ancient and, well, a little less ancient, to learn how they've lasted so long. We'll find out what scientists know, and what they don't, and we'll see some of the strangest sharks that have ever lived.

READY? LET'S GO!

MEASURING PREHISTORY: A RULER FOR THE AGES

Scientists love to sort and label things, and they've split prehistoric history into lots of different kinds of pieces. The most important ones for this book are eras and periods.

The Paleozoic era—the one before dinosaurs—had six periods: Cambrian, Ordovician, Silurian, Devonian, Carboniferous, and Permian.

The Mesozoic era—the one with dinosaurs—had three periods: Triassic, Jurassic, and Cretaceous.

The Cenozoic era—the one after dinosaurs—has three periods: Paleogene, Neogene and Quaternary. The Quaternary is where we live now.

All of these time periods were millions of years long, and sharks have been around for most of them. The first sharklike scales are from way back in the Ordovician, and the first shark teeth date from the Silurian period. But it wasn't until the Devonian that sharks really started to show up on the timeline. They haven't stopped making a splash since.

INVERTEBRATE ANIMALS, BRACHIPODS, TRILOBITES APPEAR

AMPHIBIANS APPEAR

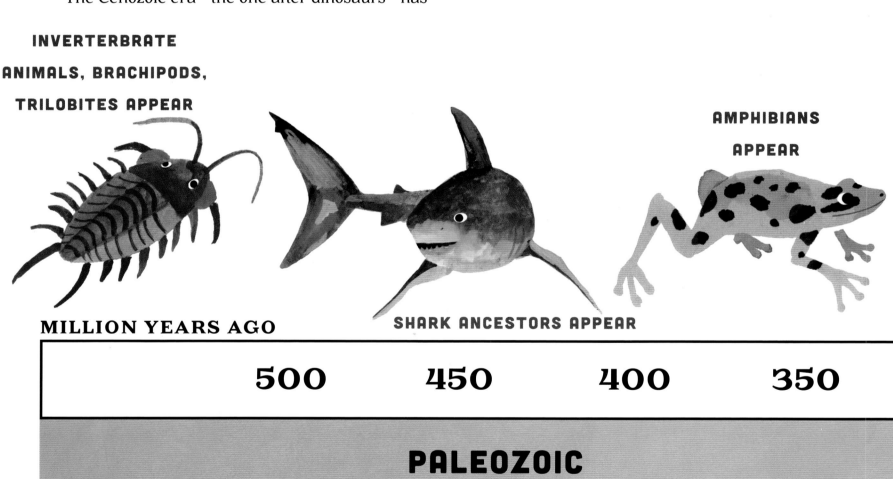

SHARK ANCESTORS APPEAR

MILLION YEARS AGO

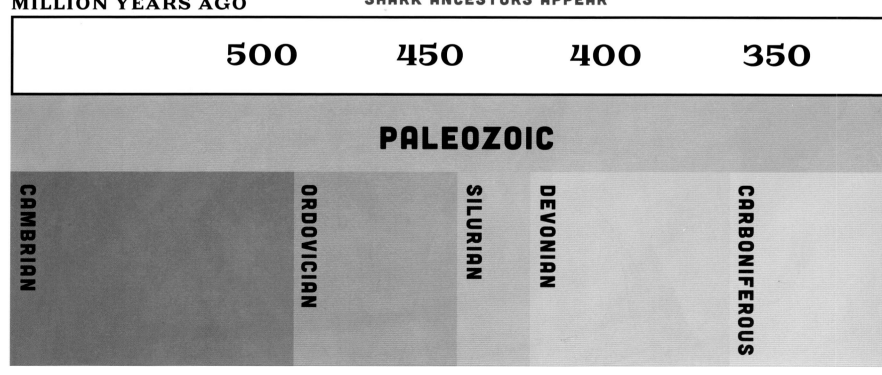

500	450	400	350

PALEOZOIC

CAMBRIAN · ORDOVICIAN · SILURIAN · DEVONIAN · CARBONIFEROUS

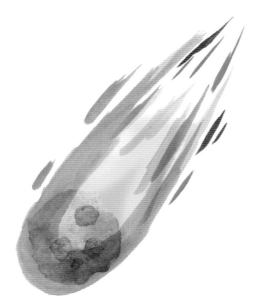

WORDS TO KNOW

We use special words to talk about prehistory. Here are some of them:

PREHISTORIC: Before there were written records.

FOSSIL: The remains of a prehistoric organism that have hardened and turned to stone.

PALEONTOLOGIST: Scientists who study fossils.

EXTINCTION EVENT: A place in the prehistoric timeline where large numbers of species went extinct. There are five great extinction events in the fossil record. The shark family survived four of them!

DINOSAURS APPEAR AND BECOME EXTINCT

HUMANS APPEAR

BIRDS APPEAR

300 250 200 150 100 50 0

MESOZOIC CENOZOIC

PERMIAN TRIASSIC JURASSIC CRETACEOUS PALEOGENE NEOGENE QUATERNARY

CLADOSELACHE

SIZE: Up to 6 feet long

ENVIRONMENT: Shallow seas

FOUND IN: North America

WHEN: 385 to 359 million years ago, Devonian period

WHAT IT ATE: Shrimp, fish, eels, other sharks

The Cladoselache lived back in the Devonian period, and like all sharks, it was also a fish. In fact, it looked like a fish dressed up as a shark for Halloween. But paleontologists think Cladoselache might have been one of the earliest true sharks. It had hardly any scales, and its mouth was at the front of its snout instead of below it. But Cladoselache had the true body shape of a shark, making it fast and agile. Many of the prey found in Cladoselache fossils were swallowed tail first, which means they were probably caught while swimming away. Some Cladoselache fossils even have other sharks in them!

TOOLBOX

SHARK SHAPE

Most sharks are shaped like long pointy ovals, with a broad body that tapers at the head and tail. Most sharks have five stiff fins: two dorsal fins on the back, a caudal—or tail—fin, and two pectoral fins where humans would have arms. The pectoral fins are important because, like the wings of an airplane, they help keep the shark from tilting down and running headfirst into the bottom of the ocean! Sharks also have two or three smaller fins underneath their bodies that help keep them stable. That way they don't roll over in the water.

Even though most sharks are shaped basically the same way, they can still look very different from each other. Shark bodies can be sleek and stream-lined or thick and muscular. Shark snouts can be long or short, blunt or pointy. Tail fins can be thin and flexible or large and powerful. Dorsal fins can be large or small, set forward toward the head, or set farther back. Some sharks have only one dorsal fin!

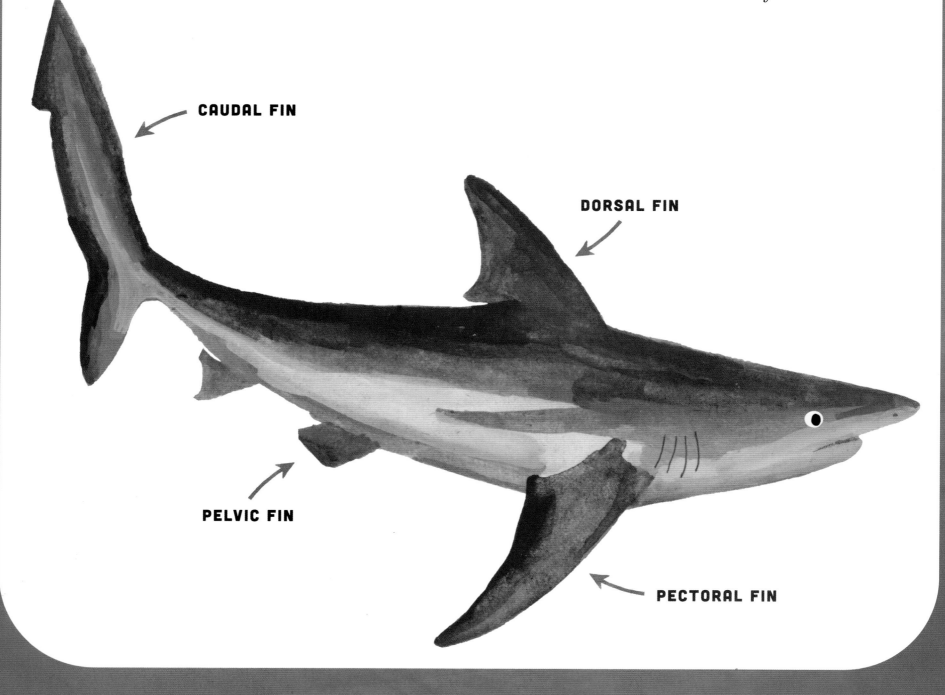

CAUDAL FIN

DORSAL FIN

PELVIC FIN

PECTORAL FIN

BEAR GULCH BAY

Close to Lewistown, Montana, is a fossil dig called the Bear Gulch Limestone. There, in the remains of what once was a tropical lagoon, scientists have found wonderfully preserved fossils. Bear Gulch holds more than 130 different species of fish, including 65 species of prehistoric sharks. When paleontologists started digging in the limestone there, they found many species they'd never seen before!

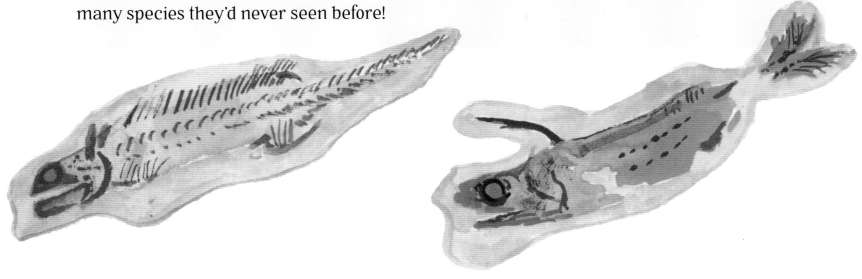

STETHACANTHUS

SIZE: 2 to 3 feet long (a few larger species grew up to 9 feet long!)

ENVIRONMENT: Warm coastal waters

FOUND IN: North America, Europe, much of Asia

WHEN: 370 to 260 million years ago, late Devonian to Permian periods

WHAT IT ATE: Small fish, crustaceans, and cephalopods

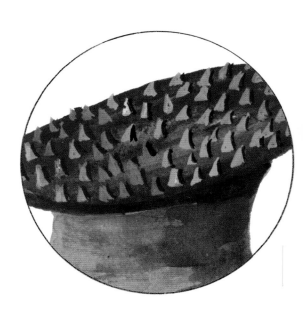

Stethacanthus is one of the most unusual sharks in history. The males had a unique, flat-topped dorsal fin that paleontologists call a spine-brush complex. The top of the dorsal fin was covered in spiky skin. It might have been used for mating or for scaring off predators. There was a matching spiky patch on the top of its head. On either side of its pectoral fins, it had long, thin, flexible whips, four in all. Stethacanthus was a *cladodont*, which means that at the base of each of its teeth were tiny tines, like on a fork. This helped the shark hold on to its food and swallow it whole.

Some kinds of Stethacanthus were large, but others were quite small, weighing little more than a cat. They probably weren't very fast either. Scientists think this shark kept close to shore, possibly feeding off the shallow ocean floor or scavenging from larger predators.

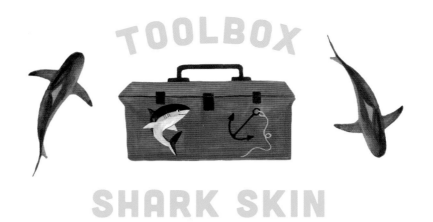

The spikes on the Stethacanthus's fin and head were a larger version of the kind of skin that sharks have today. Modern sharks have very thick skin coated with small scales called dermal denticles. These special scales are covered with enamel and often very sharp, as if the whole shark was covered with tiny teeth. In fact, *dermal denticles* literally means "skin teeth."

Dermal denticles come in all different shapes. They can be rounded or jagged. They can be shaped like teeth or thorns or leaves. The silky shark has small, tightly packed dermal denticles, which makes its skin look extra smooth. Scientists think these sandpapery scales not

only protect the sharks but also help them move more quickly through the water. The skin can also cut you badly if you rub against it the wrong way. That's one of the reasons you should never hug a shark.

SHARK SLIME

Shark skin isn't just rough; it's also slimy. Most sharks are covered with a very thin layer of mucus, although on some sharks, like bramble sharks or basking sharks, the slime is much thicker. Scientists think this slime helps protect sharks from infection and keeps things like barnacles from sticking to them.

LIFE IN THE DEVONIAN

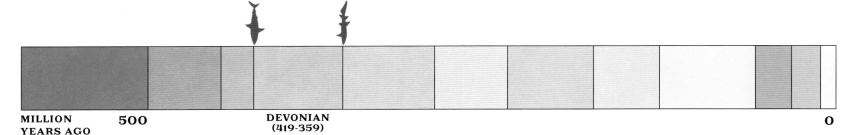

MILLION
YEARS AGO 500   DEVONIAN
(419-359)   O

Sharks and their relatives the rays made their first major appearance in the Devonian. But they were not the biggest predators.

The Devonian was the time of huge bony fishes. The 12-foot-long Hyneria ate any animal unlucky enough to have ventured into the shallow seas. It even ate animals standing at the water's edge! Dunkleosteus and its cousin Titanichthys could get up to thirty feet long. These fish had heavily armored heads and bony plates for teeth. While Titanichthys probably ate smaller things, Dunkleosteus had a bite that rivaled that of T. rex and an appetite to match. Dunkleosteus's sharp plates

and powerful jaws could crunch through the armor of other Dunkleosteus, and then the bigger fish would eat the smaller fish and spit out its bones. Sharks would have had to be careful and quick to avoid these giant predators.

On land, things were a little safer. The first forests were growing, made up of tall, fernlike trees. Insects buzzed, scorpions skittered, and large amphibians had just begun to crawl out of the water. The sharks that lived during this time would have stayed mostly in the ocean, but something new was about to be added to the toolbox.

It was almost time for freshwater sharks.

DEVONIAN EXTINCTION

Something happened at the end of the Devonian that scientists are still trying to figure out. Maybe it was a mass extinction, like the one that took out the dinosaurs. Maybe it was several smaller extinction events or a gradual decline. Whatever the reason, huge numbers of marine species did not survive into the Permian. One group that did? Sharks.

BANDRINGA

SIZE: 10 feet long

ENVIRONMENT: Swamps, river deltas, shallow coastal waters

FOUND IN: North America

WHEN: 315 to 307 million years ago, Carboniferous period

WHAT IT ATE: Fish and small marine animals

The Bandringa was a very special shark found in the rivers of what is now the midwestern United States. It had a spiny head and a long, spoonlike snout that it used to shovel around in the muddy silt of the riverbed to find and suck up small animals to eat.

The Bandringa lived in freshwater swamps and rivers, but scientists now think it traveled downstream so it could lay its eggs in the shallow ocean water of the coasts. It's one of the earliest known sharks to migrate in this way. The nurseries of the Bandringa are some of the best-preserved fossil nurseries we have, and they contain both eggs and babies. Juvenile Bandringa were only 4 to 6 inches long, but the adults could be up to 10 feet!

TOOLBOX

SHARK NURSERIES

Sharks can have babies in a few different ways. Some sharks, like hammerheads, give birth to live young, called pups. Like dogs, sharks can have litters of pups, anywhere from just two pups to as many as one hundred. The whale shark, which hatches eggs inside its body, can have up to three hundred babies!

About one-third of sharks lay eggs. Shark eggs are covered in a thick leathery egg case and can be anchored to reefs or seagrass. Some have adhesive fibers that help them stick to the seafloor, and bullhead shark eggs are shaped like a drill so they can be wedged into rock crevices. Large groups of eggs are called nurseries.

Some sharks, like the whale shark, carry their eggs inside their bodies to keep them safe. The eggs have a very thin film around them instead of a shell, and the sharks carry them until the young are ready to hatch out. Other sharks, like the sand tiger shark, will actually eat most of their other siblings before they are officially born!

If they do give birth to live young, many sharks will seek out safe, shallow water for their babies. Coastal waters make good nurseries because the water is warmer and there are lots of small animals for the young sharks to eat.

SHARK MIGRATION

Many sharks today migrate, swimming to special places to lay eggs or give birth to live young. Sharks will also migrate to mate, to find food, or to move to warmer waters as the season changes. Tiger sharks can swim more than 4,000 miles each year!

Tag location
28 Oct 2018

First transmission
1 Feb 2019

Last transmission
18 Aug 2019

Tiger shark tracked 13,000 km over 263 days

ORTHACANTHUS

SIZE: 10 feet long

ENVIRONMENT: Tropical coastal swamps and bayous

FOUND IN: Europe and North America

WHEN: 400 to 225 million years ago, early Devonian to late Permian periods

WHAT IT ATE: Fish, smaller amphibians, sometimes baby Orthacanthus

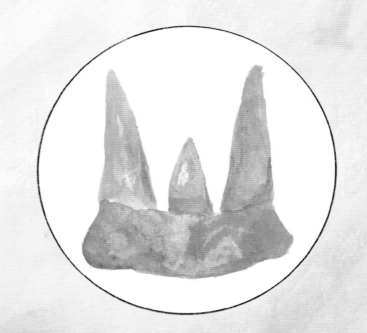

Orthacanthus was an apex—or top—predator. It was big and fierce, and it ruled the freshwater bayous of Europe and North America for more than 100 million years, along with its cousin, the Xenacanthus. Having a long, flexible body helped it navigate swamps without getting stuck in the thick vegetation. It had three-pronged, or tricuspid, teeth and a spike growing out of its skull. The word *orthacanthus* means "vertical spike." Orthacanthus ate anything it could catch, including other smaller Orthacanthus. We know this because we've found juvenile Orthacanthus teeth in fossilized adult Orthacanthus poop. How did we know it was Orthacanthus poop? Because this shark is one of the only animals we know that poops in a spiral!

TOOLBOX

FRESHWATER SHARKS

Freshwater sharks, or river sharks, are still around today. The rare Ganges shark is a true freshwater shark. It lives only in the freshwater rivers of Bangladesh and India, possibly into Pakistan and Myanmar. The northern river shark and the speartooth shark swim in the tidal coastal rivers of Papua New Guinea and Australia and are found in both salt water and fresh water. All of these species are highly endangered, and not much is known about them.

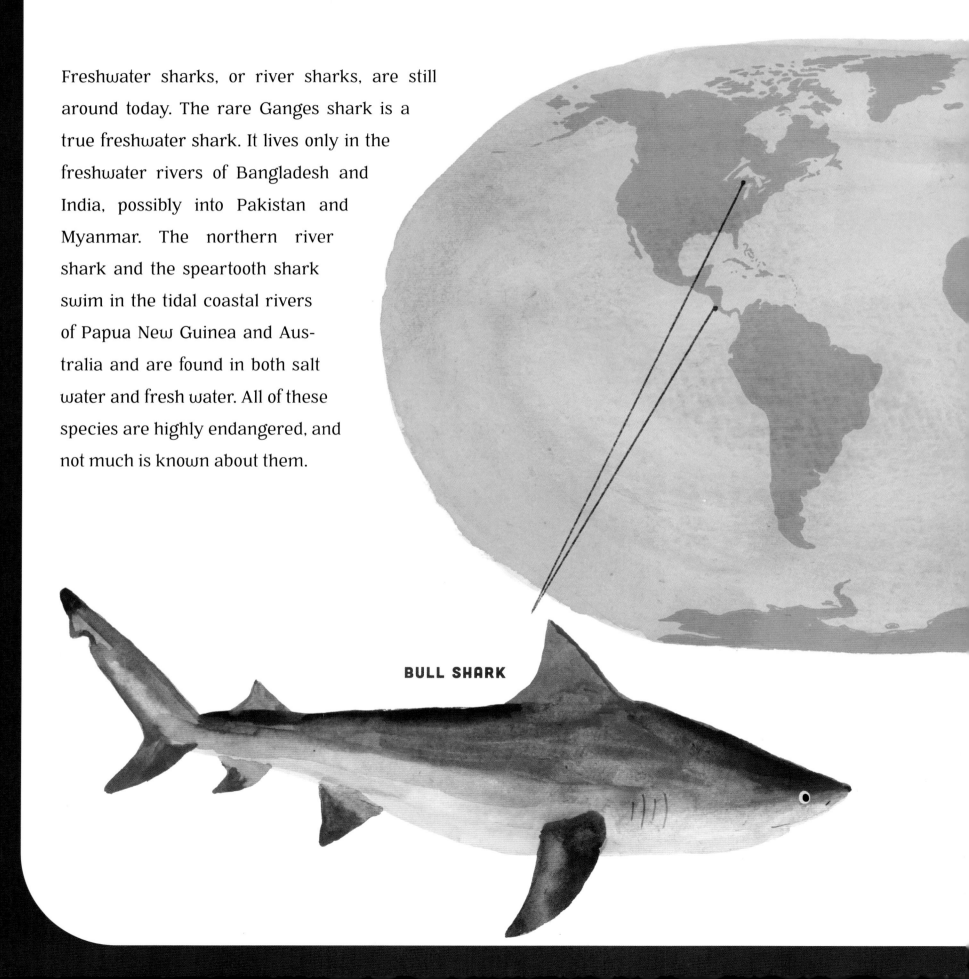

BULL SHARK

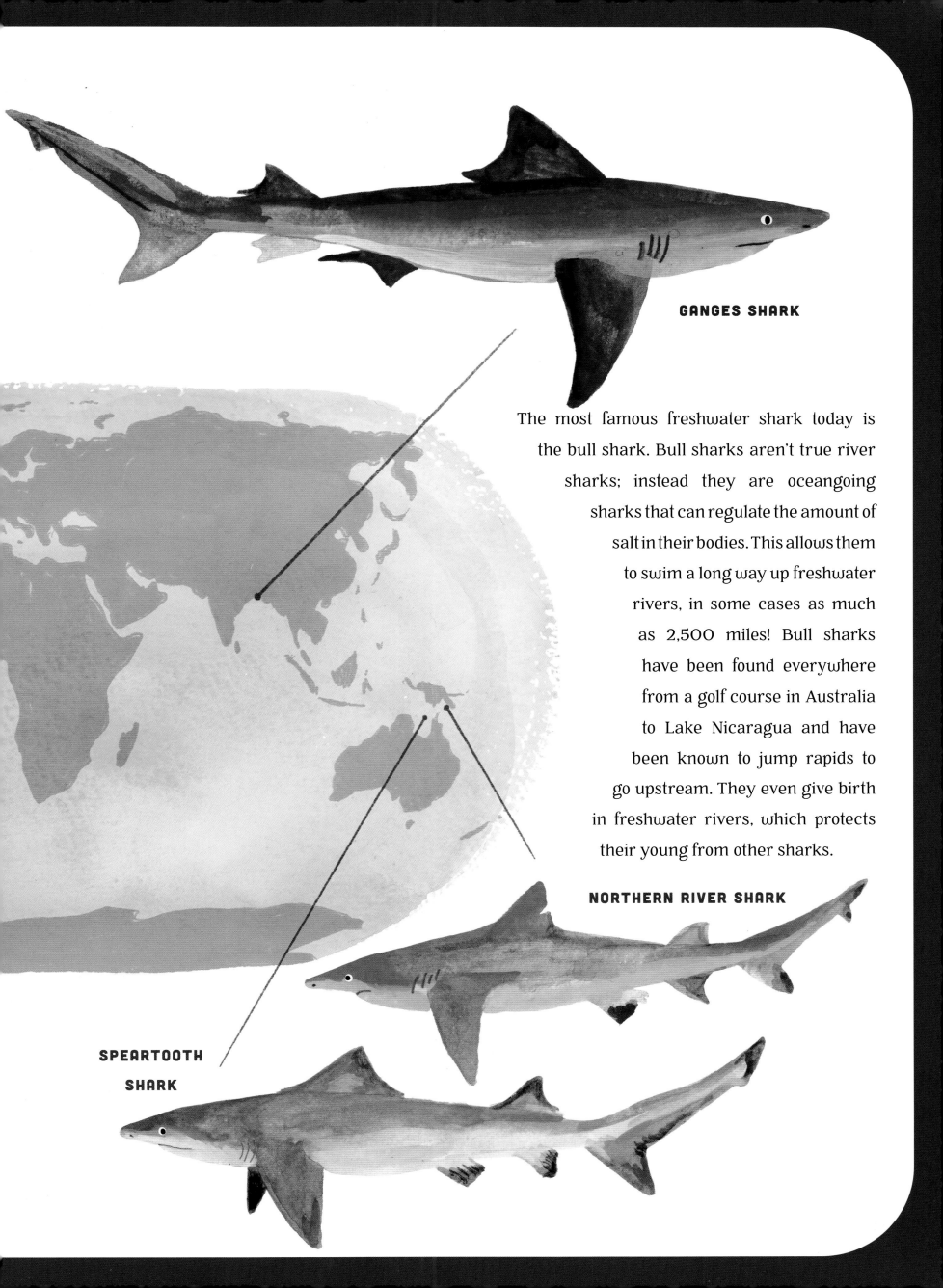

GANGES SHARK

The most famous freshwater shark today is the bull shark. Bull sharks aren't true river sharks; instead they are oceangoing sharks that can regulate the amount of salt in their bodies. This allows them to swim a long way up freshwater rivers, in some cases as much as 2,500 miles! Bull sharks have been found everywhere from a golf course in Australia to Lake Nicaragua and have been known to jump rapids to go upstream. They even give birth in freshwater rivers, which protects their young from other sharks.

NORTHERN RIVER SHARK

SPEARTOOTH SHARK

JUST KEEP SWIMMING

Sharks have gills on the sides of their heads, which means they can take oxygen out of the water. But for sharks to breathe, water has to keep flowing over the gills. This is why some sharks, like hammerheads, great whites, and whale sharks, must keep swimming all the time. If they stop moving, they won't be able to take oxygen from the water. But, then, how do they sleep? Scientists aren't sure. We do know that sharks don't sleep like humans do and instead rest with their eyes open. Sharks like the great white may do so in places where the current is strong enough to keep the water flowing.

Some sharks *can* lie still. Sharks like angel sharks, whitetip reef sharks, and nurse sharks rest on the bottom of the sea and take in water through small openings set behind their eyes, called spiracles. Other sharks will gulp water, which keeps it flowing over the gills. They can even suck up food this way!

WHAT POOP CAN TELL US

Coprolite, what we call fossilized animal poop, is an important scientific way to see what an animal eats and often where and when they ate it. Some scientists spend their lives just studying these dung stones!

Spiral Orthacanthus poop like this was found near New Brunswick, Canada, and contained teeth from smaller, juvenile Orthacanthus.

This ancient poop, possibly from a crocodile ancestor, has bite marks and a shark tooth in it. Scientists think the shark mistook the poop for a tasty meal. Yuck!

This tiny (1.5-inch) fossil poop was found in South Carolina and belonged to a newborn shark. The bones inside it are from a baby turtle!

LIFE IN THE CARBONIFEROUS

MILLION YEARS AGO   500   CARBONIFEROUS (359.2–299)   0

By the time the Carboniferous happened, the armored fishes had died out, leaving a world where sharks were top predators. Sixty percent of fish back then were sharks or shark relatives. Early squid and octopus appeared, and amphibians grew to the size of alligators. But sharks ruled them all, at least in the water.

On land, it was a different story. Hiding in the thick forests of ferns was a world of giant bugs.

Back then, the air was warm, with much more oxygen than we have today, which helped insects grow to huge lengths. There were mayflies the size of parakeets, dragonflies the size of hawks, and scorpions that grew as big as beagles. The Arthropleura, a relative of the millipede, could grow to more than seven feet long! As the land masses began to come together, the climate changed, and the ocean levels began to fluctuate. This turned the shallow seas into swamps and deltas. At the end of the Carboniferous, many of the forests collapsed and fell into the rivers and estuaries, creating the peat bogs that would turn into coal over millions of years. When the trees began to die, the giant insects died as well, leaving only their smaller descendants behind.

HELICOPRION

SIZE: 20 to 30 feet long or more
ENVIRONMENT: Open ocean
FOUND: All over the world
WHEN: 290 to 250 million years ago,
Permian to Triassic periods
WHAT IT ATE: Soft, deep-sea prey like squid
or other sharks

Helicoprion was one of the strangest creatures in the prehistoric seas. A relative of sharks, all we could find for a long time were Helicoprion's teeth. But what teeth they were! Up to 180 teeth were set in a round spiral pattern the size of a dinner plate. This unique fossil, called a tooth whirl, puzzled scientist for more than a hundred years. No one knew how those teeth fit into the shark's mouth.

Scientists have used new technology to scan fossil Helicoprion jaws. We now know that the tooth whirl was set back inside the mouth, and they were the only teeth the creature had. We also know from the tooth wear patterns that this shark probably ate mostly soft, non-bony things, like squid and other sharks. It's possible that Helicoprion charged its prey with its mouth open and sliced it in half!

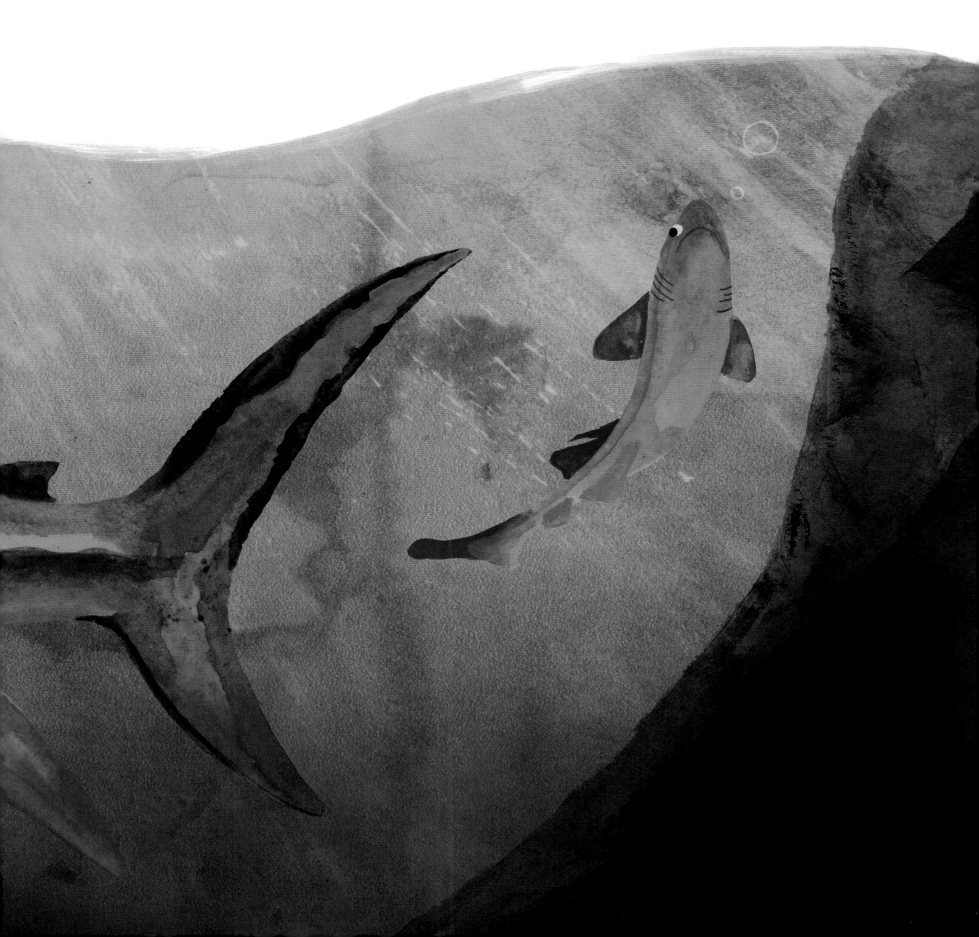

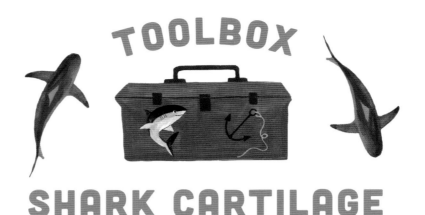

TOOLBOX

SHARK CARTILAGE

So why did we have only the Helicoprion teeth? The reason has to do with the shark's skeleton. Sharks are vertebrates, which means they have skeletons on the inside, like humans, instead of armor on the outside, like bugs. But shark skeletons are made of cartilage, not bone. That is why sharks and their relatives are called cartilaginous fish. Cartilage is a kind of firm tissue, the sort that makes up your ears and nose. It's lighter than bone, which lets the sharks move faster and with less effort. But it also breaks down faster

than bone, and so it doesn't fossilize well. Some sharks have spines in their fins that fossilize, and in some places, like Bear Gulch Bay or the rivers where the Bandringa lived, sharks were buried in silt and preserved whole. But a lot of the time, all that scientists find are teeth.

GREAT WHITE SHARK

SHARK COUSINS

One of the surprises scientists discovered when they learned more about Helicoprion was that even though we thought it was a shark for a long time, it's not a shark at all! It's closer to a shark relative called a chimera.

Today, chimera are mostly small, bottom-dwelling fish that use thick plates to grind their food, but back in prehistory, there were many, many forms of chimera. Some were sharklike, with teeth, like Helicoprion, while others were more like tropical fish with beaks, like Belantsea. There was even one called Iniopteryx that might have soared like a flying fish. Some think Stethacanthus might have been a chimera too!

INIOPTERYX

CHIMERA MONSTROSA

BELANTSEA

THE PERMIAN EXTINCTION

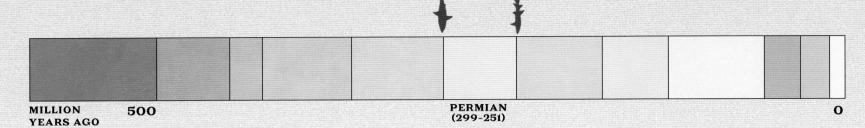

MILLION YEARS AGO   500   PERMIAN (299–251)   0

The Permian period came after the Carboniferous, and it was a time of significant climate change. The air dried out and its oxygen content became much lower. The land was full of huge reptiles, not related to dinosaurs, including the famous Dimetrodon. Shallow coastal waters became rarer as the continents came together in one large mass called Pangaea.

Seasonal extremes became harsher, and at the end of the Permian period, there was an extinction event similar to what happened when the dinosaurs died out. Only this one was worse! More than 70 percent of land animals and 95 percent of marine creatures died out. The Helicoprion was one of the survivors, but most sharks were not so lucky.

The age of sharks was well and truly over.

HYBODUS

SIZE: 6.5 feet long
ENVIRONMENT: Shallow seas
FOUND: All over the world
WHEN: 260 to 75 million years ago,
Permian to Cretaceous periods
WHAT IT ATE: Fish, various marine reptiles
and other animals

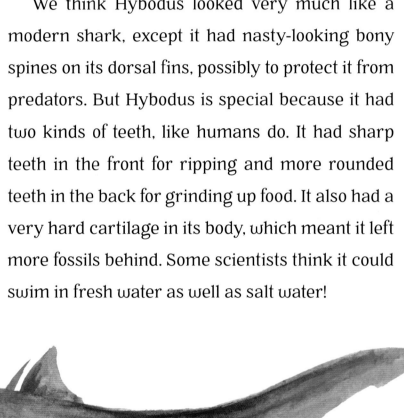

Hybodus wins the prize for top survivor in the shark world. It survived not only the Permian extinction and a smaller extinction event at the end of the Triassic period, but it also made it all the way into the Cretaceous, the time when the Tyrannosaurs lived. That's 200 million years of success! This shark was fast and strong, about the length of a tall human, and it lived all over the world.

We think Hybodus looked very much like a modern shark, except it had nasty-looking bony spines on its dorsal fins, possibly to protect it from predators. But Hybodus is special because it had two kinds of teeth, like humans do. It had sharp teeth in the front for ripping and more rounded teeth in the back for grinding up food. It also had a very hard cartilage in its body, which meant it left more fossils behind. Some scientists think it could swim in fresh water as well as salt water!

TOOLBOX
SHARK TEETH

The kind of teeth a shark has depends on what it eats. Some sharks have dense flat teeth that they use to crush the shells of animals at the bottom of the ocean. Some sharks have sharp, needle-thin teeth, good for grabbing onto slippery fish and squid. The great white shark, like the megalodon, has triangular teeth that are pointed on the bottom, with serrated edges for cutting up bigger prey. The Port Jackson shark has two kinds of teeth, like Hybodus, although it uses its teeth to crush and eat starfish and mollusks, not the faster fish and animals that Hybodus probably ate.

Sharks also have a *lot* of teeth, lined up in rows inside their mouths. The teeth work something like a conveyer belt. Each new row of teeth grows in and moves forward, pushing out the rows of teeth in front of it. Sharks lose about one tooth a week, which can add up to thousands over a lifetime. That's why we have so many shark teeth and fossils today!

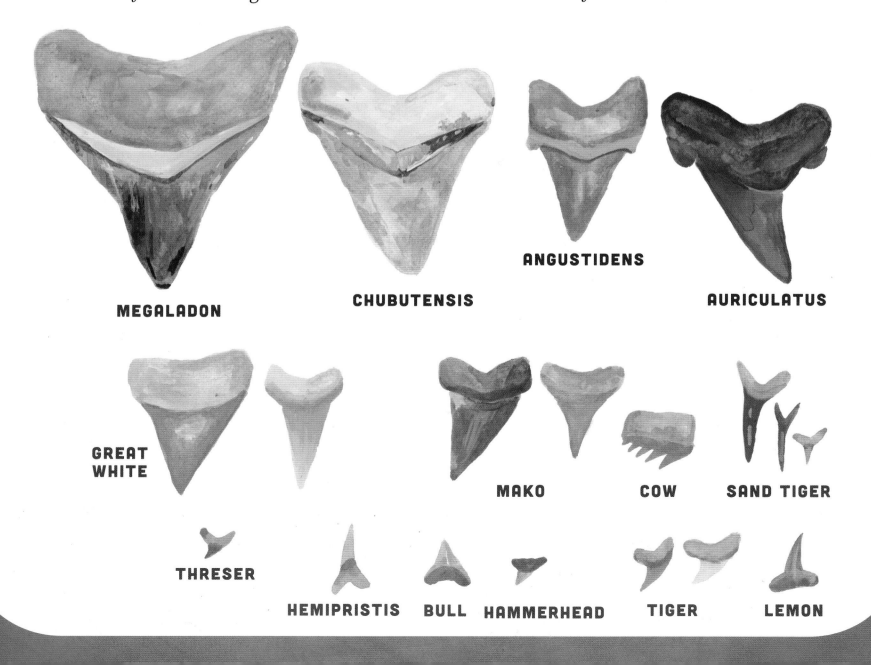

MEGALADON

CHUBUTENSIS

ANGUSTIDENS

AURICULATUS

GREAT WHITE

MAKO

COW

SAND TIGER

THRESER

HEMIPRISTIS

BULL

HAMMERHEAD

TIGER

LEMON

MARINE REPTILES

Hybodus had to be strong and fast because, unlike in the Carboniferous period, sharks in the time of the dinosaurs were not the biggest predators around. That distinction belonged to the marine reptiles of the Jurassic and Cretaceous.

Marine reptiles are often classed with dinosaurs, but they aren't actually related. Just as dinosaurs ruled the land, marine reptiles ruled the seas. Some had long necks, some were more fish shaped, and some even lived partly on land, like seals. And some were predators the size of whales. The Hybodus was probably prey for such giants as the Liopleurodon in the Jurassic and the Mosasaur in the Cretaceous. These reptiles would jump from the water to catch pterosaurs, wait in ambush for sharks, and eat anything from turtles and fish to actual dinosaurs! No wonder the Hybodus needed spikes for protection.

LIOPLEURODON

HYBODUS

SCAPANORHYNCHUS

SIZE: 2 feet long (although some could have been much larger)
ENVIRONMENT: Deep water
FOUND: All over the world
WHEN: 125 to 66 million years ago, Cretaceous period (maybe even longer!)
WHAT IT ATE: Fish and squid

Despite the fact that its name sounds like a sneeze, the Scapanorhynchus was a serious shark. It had a long, flat, pointy snout—like a sword—and lots of sharp, needlelike teeth that it used to hang on to slippery prey. Like its relative, today's goblin shark, it lurked down in the deep waters of the ocean bottom, waiting to munch on fish and squid. It had a much larger tail fin than the goblin shark though and was generally smaller. Most Scapanorhynchus were only about two feet long, but goblin sharks today can get up to 20 feet!

TOOLBOX
SHARK JAWS

Scapanorhynchus likely shared something else with goblin sharks: loosely attached jaws that could be extended like a grabber claw to snatch prey. A lot of sharks, including the great white, have jaws like these, but the goblin shark's jaws can reach the farthest. They can also open their mouths 111 degrees, much wider than a human mouth! Like many sharks, they swim with their mouths open, in order to breathe. When prey gets near, they snap their jaws shut.

To help strengthen their jaws, sharks have a thin covering of tiny, six-sided, calcium salt crystals called tesserae over the cartilage that holds their jaws together. (Calcium salt is one of the things that makes our own bones stronger.) This special bite gives the shark more power and helps it create suction to bring food into the mouth. Goblin sharks can bite faster than a cobra can!

HOW SHARKS FLOAT

Sharks, like most fish, are heavier than water, but most sharks don't have swim bladders to keep them from sinking. Instead, many sharks have large livers full of a very light oil that helps them stay buoyant in the water. The blue shark uses its extra-large pectoral fins to keep it afloat, while sand tiger sharks will surface and gulp air into their stomachs. Then the sharks will fart it out!

THE CRETACEOUS EXTINCTION

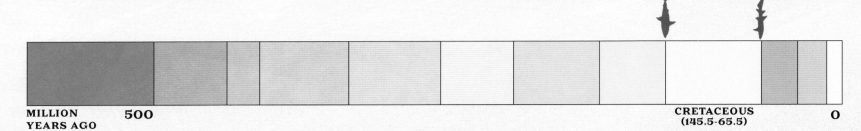

MILLION
YEARS AGO   500   CRETACEOUS   0
                                                                               (145.5-65.5)

The age of the dinosaurs came to an end with the famous Cretaceous extinction event. Most scientists think a comet hit the earth, setting off volcanoes and earthquakes. The impact and the erupting volcanoes threw lots of ash and acid into the air, blocking the sun. The climate cooled too quickly for the animals to adapt, and 75 percent of the plants and animals died off.

More than anything, size determined who would live and who would die. With a few exceptions, like turtles and crocodiles, no vertebrates larger than 55 pounds survived. This included amphibians, dinosaurs, and many, many sharks. Birds and mammals were still small, so in the end, they were the ones who inherited the new empty earth.

The remaining sharks would make a comeback though . . . and get even bigger.

DRAVIDOSAURUS

EUTHERIA

JUBBULPURIA

JAINOSAURUS

MEGALODON

SIZE: 50 to 60 feet long

ENVIRONMENT: Warm and coastal waters

FOUND: All over the world

WHEN: 23 to 2.6 million years ago, Neogene period

WHAT IT ATE: Anything it wanted to (mostly small whales)

Megalodon means "big tooth," and it is probably the best-known prehistoric shark of all. It was bigger than a city bus and weighed up to 100 tons. It had huge, serrated teeth twice as long as a great white shark's teeth, with ten times the bite force. Megalodon might have hunted by biting the fins and tails off whales and large fish, making sure its prey couldn't escape. Some people think megalodons might still be alive, but the truth is they are very, very extinct. Megalodon was not a deepwater shark; it stayed near the surface and swam along the continental shelf near the coasts. It would be hard to miss a megalodon if it were still around!

SEA OF GIANTS

Megalodon was not the only giant in the seas. It often had competition from huge predatory whales the size of . . . well . . . whales! The leviathan whale, for example, had teeth up to fourteen inches long, the largest interior teeth of any known animal. And before the megalodon existed, the ruler of the seas was Basilosaurus, a whale that used its long narrow shape for deadly speed. Because of these hunting giants, the seas of the Paleozoic and the Cenozoic were some of the most dangerous ever known.

TOOLBOX

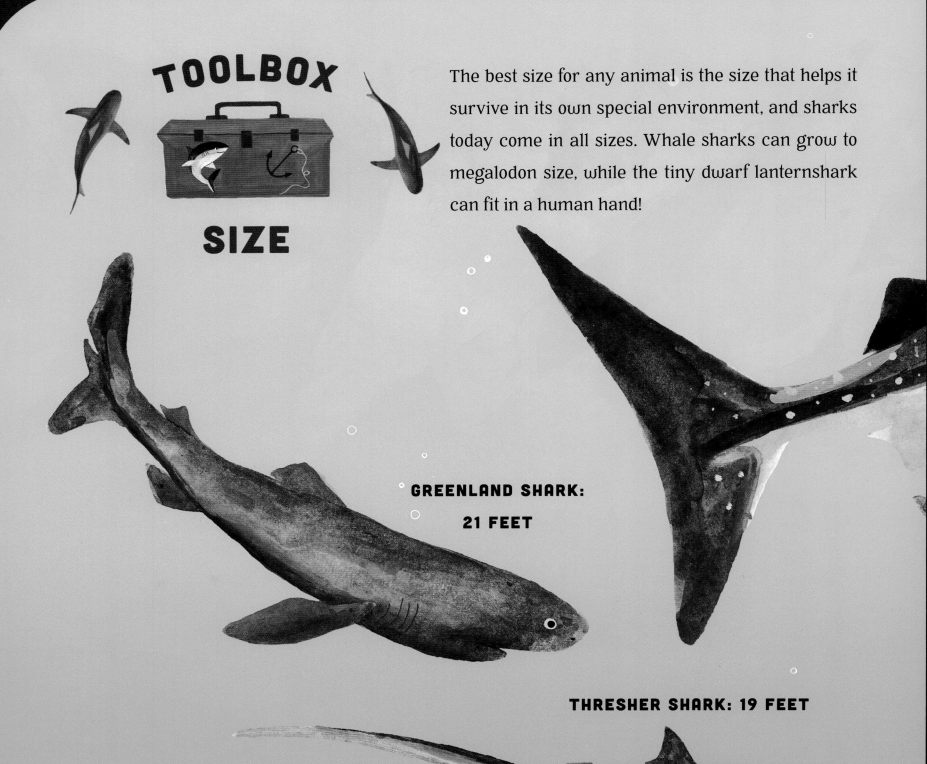

SIZE

The best size for any animal is the size that helps it survive in its own special environment, and sharks today come in all sizes. Whale sharks can grow to megalodon size, while the tiny dwarf lanternshark can fit in a human hand!

GREENLAND SHARK: 21 FEET

THRESHER SHARK: 19 FEET

BULL SHARK: 11 FEET

WOBBEGONG SHARK: 10 FEET

SAWSHARK: 4.5 FEET

FRILLED SHARK: 7 FEET

BONNETHEAD SHARK: 5 FEET

COOKIECUTTER SHARK: 1.6 FEET

DWARF LANTERNSHARK: 8 INCHES

WHALE SHARK: 61 FEET

BASKING SHARK: 40 FEET

GREAT HAMMERHEAD SHARK: 20 FEET

GREAT WHITE SHARK: 23 FEET

TIGER SHARK: 24 FEET

GOBLIN SHARK: 12.5 FEET

SAWSHARK

SIZE: Up to 5 feet long
ENVIRONMENT: Tropical waters, mostly near the continental shelf
FOUND: All over the world
WHEN: 100 million years ago, Cretaceous period through the present day
WHAT IT EATS: Small fish, squid, and crustaceans

The first sawsharks date from the Cretaceous period, but they're still around today, along with a relative, a kind of ray called the sawfish. Sawsharks are unique because of their long snouts, which are edged with thin, sharp teeth. They swim in schools and have been found in tropical oceans all over the world. Like most sharks, they use their snouts to sense electrical fields from their prey, like fish swimming by or crustaceans buried in the sand. But unlike other sharks, the sawshark can actually use its nose to attack its prey, sometimes cutting them in half!

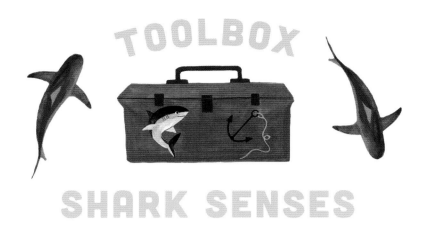

TOOLBOX
SHARK SENSES

Sharks' senses are part of what make them such good hunters. They can hear low frequency sounds that humans can't, and they have a keen sense of smell, which they use to detect blood from a long way away. Like humans, some sharks can see in color, and like cats, they have excellent night vision.

But they also have two senses that humans don't. One is a special sense called electroreception. Every creature gives out a weak electrical field when its muscles move, and all sharks can sense that field. In fact, sharks are more sensitive to electricity than pretty much any other animal! This is because they have tiny pockets of gel on their noses with special sensors, and those organs can pick up things like heartbeats and

muscles moving. They may even be able to detect magnetic fields. Sharks, like all fish, have a row of special cells along their body called the lateral line. These cells have tiny hairs that detect vibrations in the water. The lateral line and electroreception mean that sharks are extremely good at hunting in the dark, because they can sense fish even if they can't see them.

Sharks have taste buds too, like we do, but they keep them behind their teeth, not on their tongue. This means they have to bite something in order to find out whether it's food or not. Think of them as giant, sharp puppies!

OCEAN LAYERS

The ocean has five layers, each one deeper and darker than the last. The top layer is the sunlight zone, where animals and plants can both live and where there is lots of light. The second layer is the twilight zone, where light is scarce and plants can't grow. The third one is the midnight zone, where there is no light at all. The deepest layers are the abyss, or the very bottom of the sea, and the trenches, which are like canyons underwater. As far as we know, sharks live only as far down as the midnight zone, but some sharks have been known to dive deeper. And there's a lot of deep ocean left to explore.

WOBBEGONG SHARKS

SIZE: Up to 4 feet, some species up to 10 feet in length
ENVIRONMENT: Shallow temperate and tropical waters
FOUND IN: Western Pacific Ocean and eastern Indian Ocean
WHEN: Unknown through the present day
WHAT IT EATS: Crustaceans, octopus, other fish

At first glance, the wobbegong shark doesn't look much like a shark at all. It has a flattened shape and a short tail. Instead of being mostly one color, wobbegongs are covered in intricate patterns of lines and splotches and spots. They also have whiskery tassels at the front of their heads. Because of their odd appearance, this family of sharks is often referred to as carpet sharks.

Looking like a rocky carpet is very useful for wobbegongs, because they are ambush predators. That means that instead of swimming around looking for prey, they hide on the bottom of the seafloor and wait for fish to swim close so they can swallow them up. Sometimes, wobbegongs will sneak up on prey from a distance!

Because they spend so much time lying on the sand, wobbegongs don't breathe through their mouths much at all. Instead, they breathe through large spiracles, gill slits set close to the eyes.

TOOLBOX

CAMOUFLAGE

Wobbegongs, and other flat sharks like angelsharks, rely on camouflage. This camouflage isn't about *changing* color, though. Camouflage is how we refer to the way many animals are patterned to blend into their environment. The wobbegong's coloring matches the rocky reefs it lives on. Some sharks have faint spots or stripes that look like a speckled seafloor, and the lemon shark, which hunts over pale sandy beaches, is actually yellow!

Even sharks that are simply gray or brownish use color to their advantage. These sharks have something called countershading, which is when an animal is darker on top and lighter underneath. This is a kind of camouflage too! If a shark was all one color, sunlight would light up the top half of the shark, and it would be easy to see the recognizable shape of the fin. But when the shark is darker on top, light from above makes it all one shade, like a shadow. This also means that when you are under a shark and looking up, the lighter belly matches the sunlight, making it harder to see from below. This kind of shading is how sharks can hide in open water.

LEOPARD SHARKS

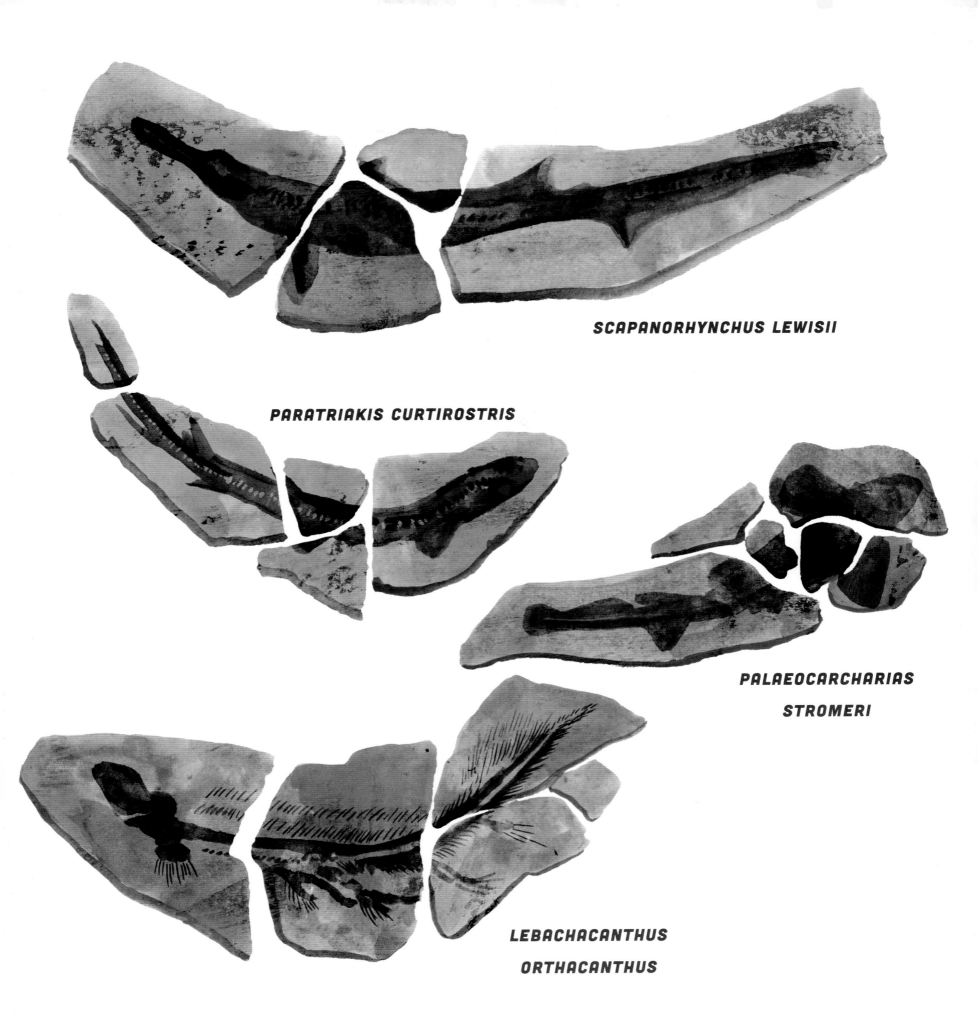

SCAPANORHYNCHUS LEWISII

PARATRIAKIS CURTIROSTRIS

PALAEOCARCHARIAS STROMERI

LEBACHACANTHUS ORTHACANTHUS

FOSSIL PUZZLES

Why don't we know how old the wobbegongs are? Well, because the fossil record is like a giant puzzle and sometimes there are missing pieces. The oldest modern wobbegongs are around 6 to 10 million years old, and fossils of relatives of the wobbegongs have been found as far back as the Cretaceous. But scientists don't have enough of a fossil record for wobbegongs right now to say for sure when these specific sharks appeared.

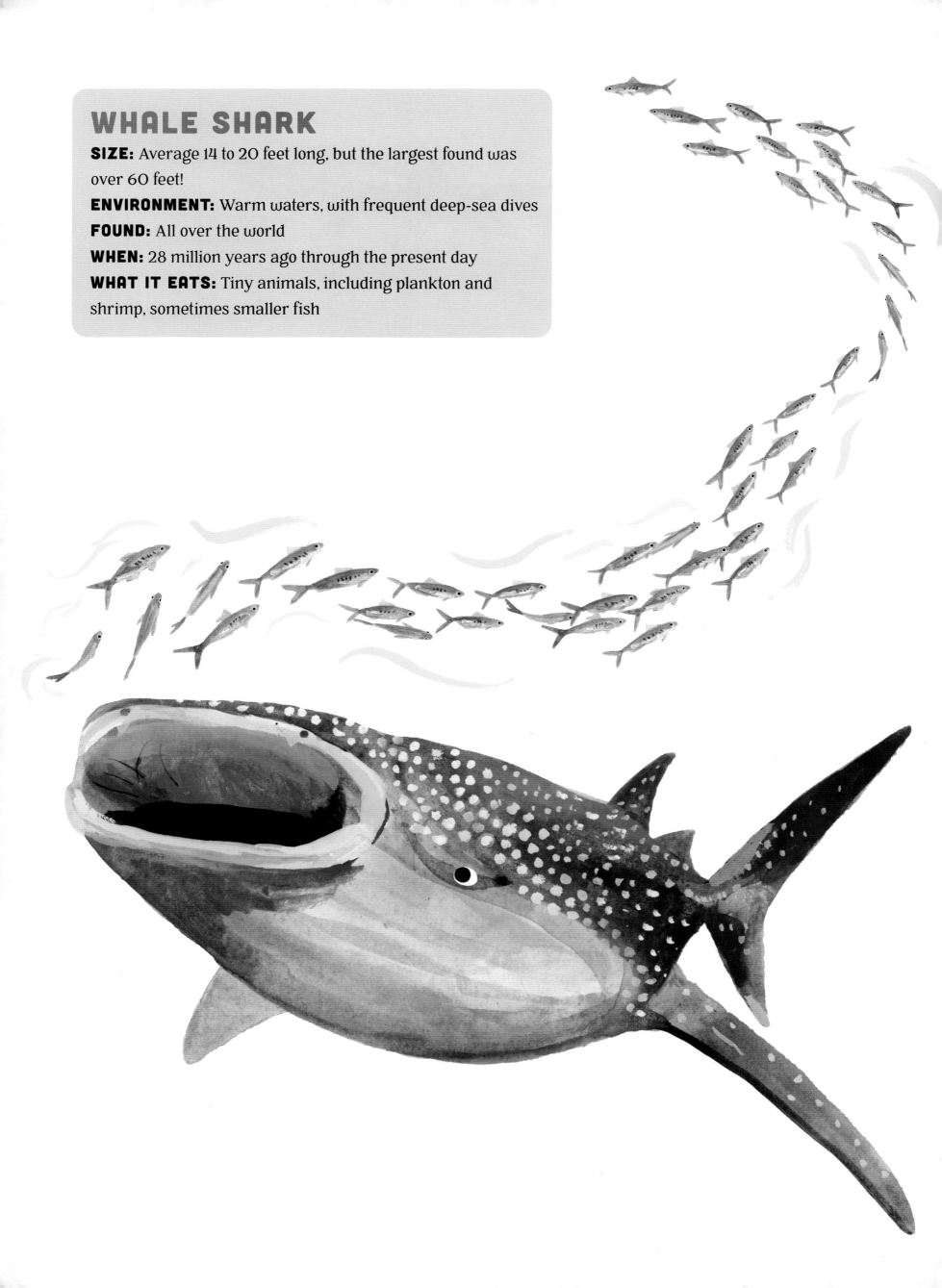

WHALE SHARK

SIZE: Average 14 to 20 feet long, but the largest found was over 60 feet!

ENVIRONMENT: Warm waters, with frequent deep-sea dives

FOUND: All over the world

WHEN: 28 million years ago through the present day

WHAT IT EATS: Tiny animals, including plankton and shrimp, sometimes smaller fish

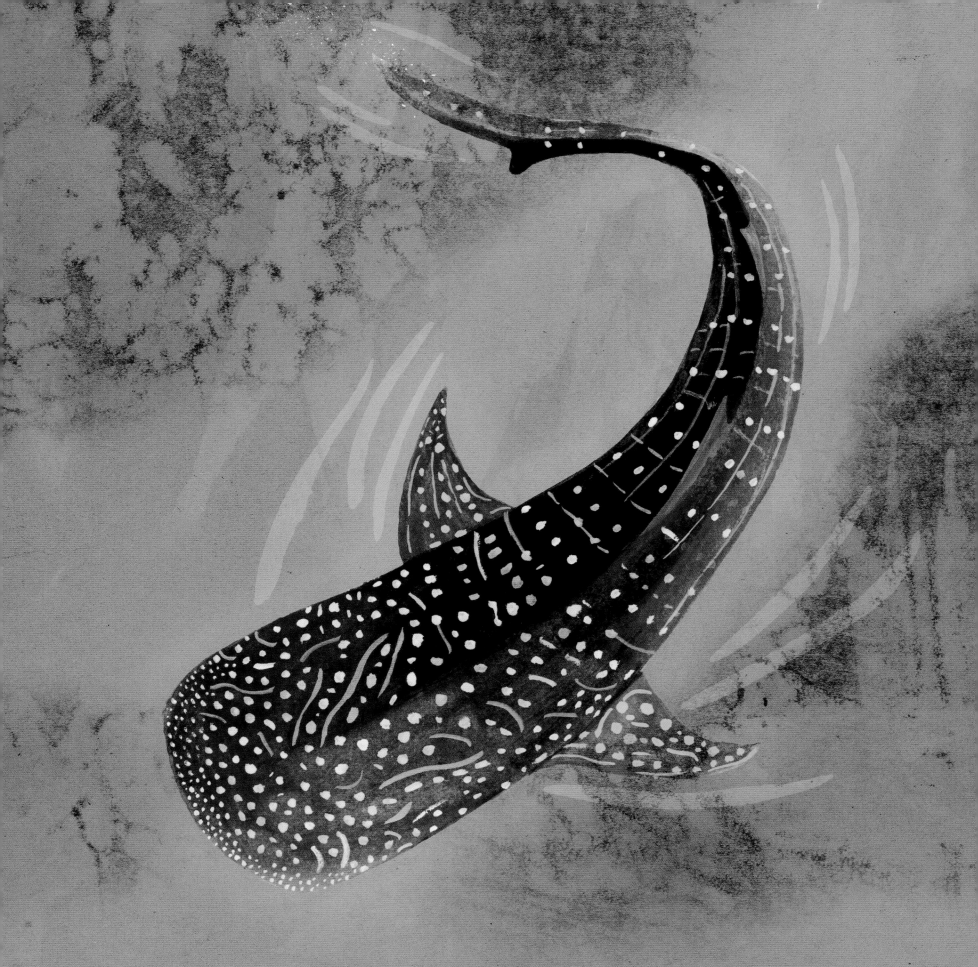

The whale shark is the biggest fish in the oceans today and can grow to the size of a megalodon. But unlike megalodon, whale sharks are filter feeders, which means they eat only tiny animals they filter from the water. They don't hunt; instead, they open their mouths very wide and swim forward, catching food along the way. Sometimes they stay still and suction water into their mouths like a vacuum cleaner.

Whale sharks have skin that is up to four inches thick, more than 300 rows of tiny teeth, a mouth at the front of their heads instead of underneath, and the ability to dive down as much as 6,000 feet. They also have spots on their backs, which some scientists think might be a kind of camouflage, to blend in with the sunlight patterns on the water. Why does a shark that doesn't hunt need to blend into the surroundings? No one knows, yet.

TOOLBOX
FILTER FEEDING

There are a lot of filter-feeding giants in the sea today. The family of baleen whales, including the gigantic blue whale, also eats very small animals.

But whales and sharks filter feed in very different ways. Unlike baleen whales, sharks don't strain food through long thin teeth plates. Instead, they open their mouths very wide so water flows into the mouth and over the gills. Small food gets caught in special gill rakers, which are lined with soft filter pads, so the shark can strain out food and swallow it.

While the whale sharks are active filter feeders, basking sharks are much more relaxed about feeding. They just keep their mouths open all the time and let the water flow in as they swim. The rare deep-sea megamouth shark is a passive filter feeder too, and it swims very slowly while taking in water. Its mouth is lined with bioluminescent organs that attract small fish and other prey.

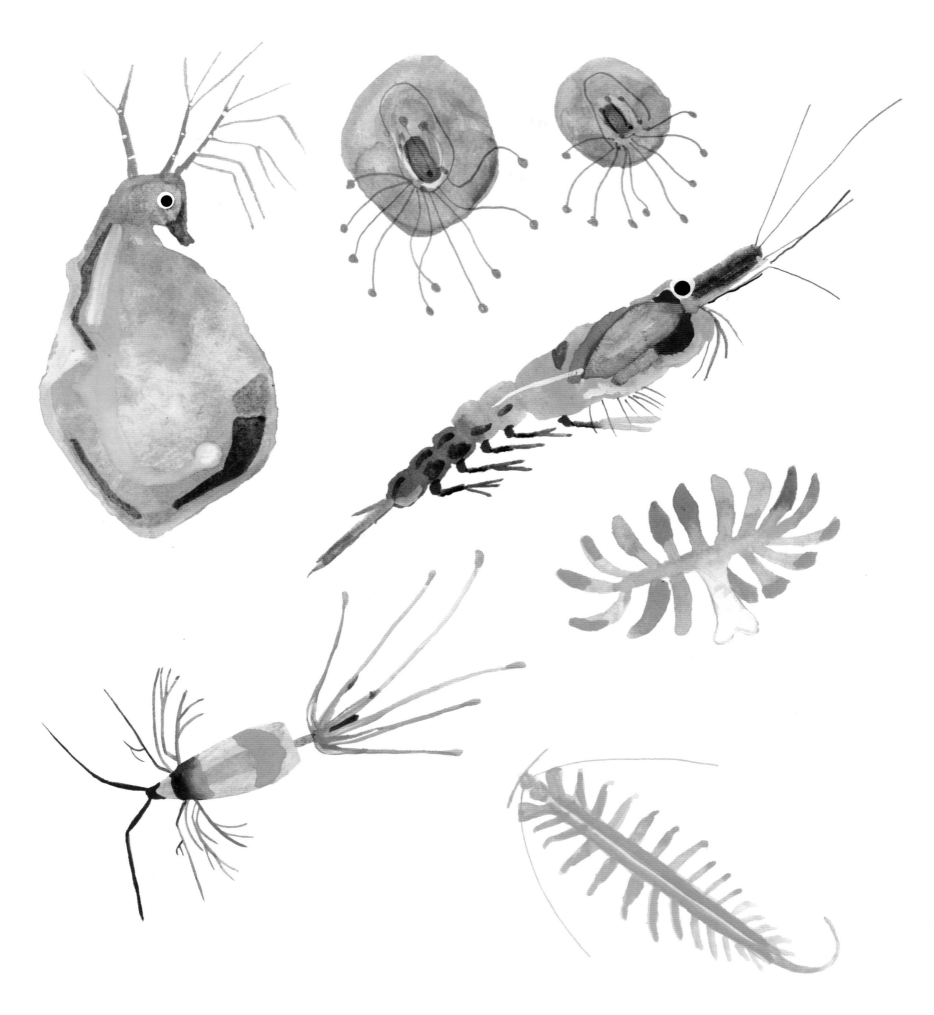

ZOOPLANKTON

The food that filter feeders eat is called zooplankton. These are tiny, tiny creatures that float in water, like krill, worms, fish eggs, and animal larvae (which are young creatures that have just hatched, like starfish and tuna.) Zooplankton drift along with the ocean currents. They can eat the algae and bacteria that bloom on the surface or the tiny bits of plants and animals that drift down into the deep water. They are very hard to see, but these little creatures are one of the major building blocks of life in the ocean and an important food source for fish and other animals—and they may even help pollinate seagrass!

COOKIECUTTER SHARK

SIZE: 18 to 20 inches long

ENVIRONMENT: Deep ocean during the day, shallower waters at night

FOUND: All over the world

WHEN: Paleocene (50 to 60 million years ago) through the present day

WHAT IT EATS: Round mouthfuls of blubber and flesh that it takes from bigger fish

Cookiecutter sharks are unique in the shark world because they feed on animals much, much bigger than they are. They're small sharks with big eyes that help them see in the darkness of the deep ocean where they live. At night, they come up to shallower waters to hunt. They have glowing patterns on their bodies that mimic smaller fish. Scientists think these patterns make other large fish and whales think the cookiecutter shark is prey. When the larger fish come closer to investigate, the cookiecutter shark attaches to them like a suction cup and takes a fast, circular bite out of them!

Cookiecutter sharks don't usually kill their prey. The sharks can cause humans trouble though, because they often get confused and bite submarines and other underwater equipment. They have sharp, gripping teeth on the top and big, slicing teeth on the bottom, which they lose all at once. Sometimes cookiecutters will eat their own teeth!

PORPOISE

TOOLBOX

SHARKS THAT GLOW

The glowing patterns that the cookiecutter shark uses to hunt are called "bioluminescence," and it's not the only shark that uses them. So do all the sharks in the lanternshark family, as well as the smalleye pygmy shark.

Bioluminescence is when an animal or plant makes its own light. The light on these sharks comes from something called "photophores," special organs on the skin that make a green or blue glow. Lots of deep-sea marine creatures use photophores, and some of them can also glow red. The photophores make light in many different ways, including using special bacteria.

Some sharks, like swell sharks and chain catsharks, have a different kind of glow called biofluorescence. Biofluorescent animals don't make their own light; instead, they have special cells that let them soak up sunlight and give it off as a different kind of light. This often makes the shark look bright green. Not all animals can see this kind of biofluorescence, so scientists think it might help the sharks see each other while staying hidden from prey and other predators.

GREENLAND SHARK

SIZE: Up to 21 feet long

ENVIRONMENT: Cold, deep waters

FOUND IN: The Arctic

WHEN: Greenland shark relatives found from the Cretaceous (60 million years ago) through the present day

WHAT IT EATS: Scavenged animals, fish, smaller sharks

The Greenland shark is very large and very slow, and they're the only shark in the world that lives in the Arctic all year round. Greenland sharks will migrate to the coldest part of the water each season, and they can dive down more than 2,000 feet. This means their blood is very cold, the same temperature as the water.

Scientists think the Greenland shark's very cold body temperature is why the shark grows extremely slowly. Growing very slowly helps the shark live a very long time. Scientists have found sharks at least 200 to 300 years old, maybe even older! That makes them both the longest-living shark and the longest-living vertebrate on earth today.

Many Greenland sharks have parasites on their eyes that make them blind, but because they are used to the dark deep, being blind probably doesn't bother them. Besides fish and smaller sharks, these sharks also eat polar bears, seals, and reindeer that fall through the ice. They've even been seen attacking swimming caribou!

TOOLBOX

WARM-BLOODED SHARKS

Most sharks are cold-blooded, which means their body temperature matches the temperature of the water around them. But some sharks can actually warm up their blood a little to help them hunt! This special family of sharks includes great whites, salmon sharks, and the shortfin mako.

These sharks are called regional endotherms. As they swim, they produce energy and heat. (Humans do too, which is why you get warm and sweaty when you run.) But instead of losing that heat into the water, these sharks have a special network of blood vessels that can redirect the heat to their swimming muscles. This keeps the muscles warm so the sharks can move quickly in cold water. Some sharks, like the mackerel shark, can also use this ability to keep their eyes and brain warm, so they can see and react very fast. But doing so uses up a lot of energy, so endothermic sharks have to hunt and eat a lot.

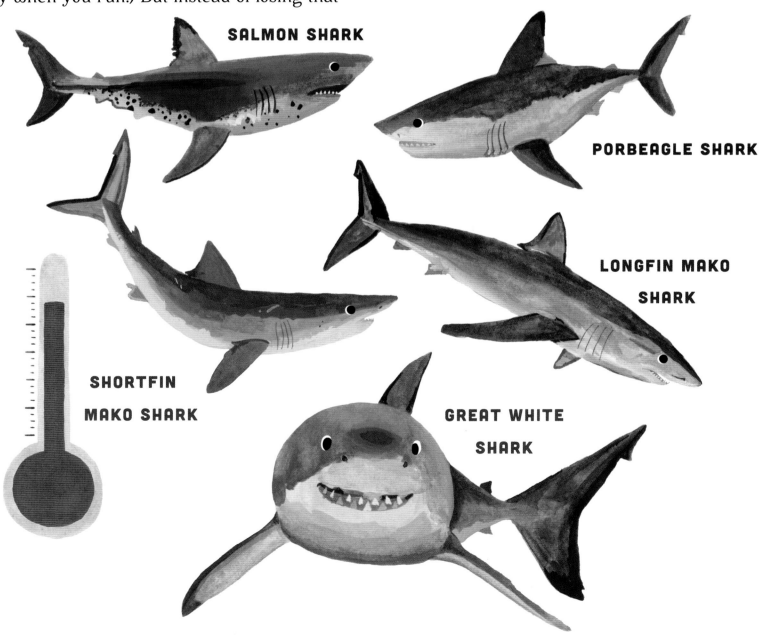

SALMON SHARK

PORBEAGLE SHARK

LONGFIN MAKO SHARK

SHORTFIN MAKO SHARK

GREAT WHITE SHARK

WARM-BLOODED OR COLD-BLOODED

Sharks and fish aren't the only creatures in the world that are cold-blooded. Reptiles and amphibians are too. These animals get hotter or colder depending on their surroundings. They can't make their own heat. This is called being ectothermic.

Humans, birds, and most mammals are homeothermic, which means our bodies are generally the same temperature all the time, no matter how hot or cold it gets outside. A lot of animals can actually do a little of both, raising their body temperature, like the endothermic sharks, or letting their bodies cool at night, like some bats.

HAMMERHEAD SHARKS

SIZE: 3 to 19 feet, depending on species

ENVIRONMENT: Warm, tropical coastlines, but migrate to cooler waters every year

FOUND: All over the world

WHEN: 20 million years ago through the present day

WHAT IT EATS: Fish, squid, octopus, stingrays, and crustaceans

Hammerhead sharks are the most easily recognized of all modern sharks, and one of the newest. But new means different things in shark time, and the hammerhead family has been around for 20 million years!

These sharks have tall dorsal fins and wide, flat heads with eyes on each end of the "hammer." This gives them the best eyesight of all the sharks. They can see in front, behind, above, and below them at all times. The broad head also helps provide lift, like an airplane wing, which means the shark can have smaller pectoral fins, allowing them to switch direction really quickly. The great hammerhead shark sometimes swims tilted to save energy and will use its flat head to pin stingrays to the ocean floor so it can eat them. Hammerheads are one of the only sharks that can get a tan!

TOOLBOX
SOCIAL SHARKS

Many people think that sharks are solitary, but not all of them are. Some sharks, like grey reef sharks, stay together and hunt in a group at night. Dusky sharks have been seen hunting whales in packs, while great white shark pairs have been known to swim together for a long time. Sand tiger sharks are social for most of the year and become solitary only during winter and spring. And of course, many sharks gather together when it's time to mate.

But hammerhead sharks are among the most social of all. While great hammerheads are solitary, others, like the scalloped hammerhead, swim in huge schools during the day. Silky sharks and basking sharks also come together in massive groups. In fact, silky and hammerhead schools can frequently be found swimming together. Some of these social gatherings can have more than a thousand sharks!

SAFETY IN THE SHALLOWS

Baby sharks can take care of themselves from the moment they are born or hatch. Their parents don't raise them. Baby sharks, like hammerheads, will stay in their nursery waters until they are old enough to strike out on their own. Sticking to the shallow nursery is good for survival, as there are fewer large predators. And while they don't really school when they're that young, it's possible the more social sharks like the company.

SCIENTISTS AND SHARKS

Scientists, like sharks, have special tools too, including imagination and persistence. They study things until they have the answer. It took 100 years to figure out how the Helicoprion tooth whirl worked. Early scientists even thought it might be part of the fins or the tail! Another tool scientists have is new technologies. That's what allowed them to scan fossil Helicoprion jaws and finally learn their secrets.

One of the most important tools that scientists have is a willingness to learn. When the fossils of the Bandringa were first discovered, scientists thought the adults and the juveniles were two different kinds of shark! It wasn't until forty-five years later, in 2014, that two scientists realized the fossils might belong to just one kind of shark, a shark that lived in the ocean as a baby and in the river as an adult. Science is always looking for new things to learn.

Today's sharks don't dominate the sea the way they used to in Carboniferous times. But they still play a vital role in the ecosystem. Today's sharks also have new threats. Warming oceans are changing sharks' normal hunting patterns, as fish and other prey move farther south and north. But it's not just the temperature that's the problem. The growing amount of carbon dioxide in the air is absorbed by the oceans and shifts the water's chemical makeup. This might have an effect on the way sharks' senses work, making it harder for them to smell food. Sharks are also greatly affected by overfishing, and by things like trophy hunting and ocean garbage. But scientists and organizations are working to protect sharks and help them thrive. And as we use our tools of imagination and exploration, we can find ways to help them too.

MORE PREHISTORIC SHARKS AND THEIR RELATIVES

ACANTHODIAN: Also called spiny sharks, these little fish were one of the first vertebrate animals to have a jaw! They had features of both sharks and bony fish, and fossils have been found as early as the Silurian period, over 400 million years ago.

COBELODUS: This unusual shark relative lived in the middle to late Carboniferous. It had a round head, big eyes, and a pectoral fin whip, like Stethacanthus. Research indicates it was probably a deep-sea hunter.

CRETOXYRHINA: This 23-foot-long shark had extremely sharp, extremely hard teeth. They most likely hunted mosasaurs and plesiosaurs in the Cretaceous. Scientists have even found a Cretoxyrhina tooth wedged in the neck of a flying reptile called a Pteranodon!

EDESTUS: Like Helicoprion, the teeth of Edestus are a mystery. This Carboniferous predator had two curved rows of teeth that sat like scissors in its mouth. Scientists think it might have moved its head up and down to slash soft prey in two!

FALCATUS: This foot-long shark lived during the early Carboniferous and its fossils are found in the Bear Gulch dig in Montana. Falcatus are unique because of their large eyes and because of the male's strange, forward facing dorsal spine.

FERROMIRUM OUKHERBOUCHI: In 2020, scientists were able to study a new Devonian shark from Morocco. This shark had a hinged jaw that allowed it to drop the sides of its jaw

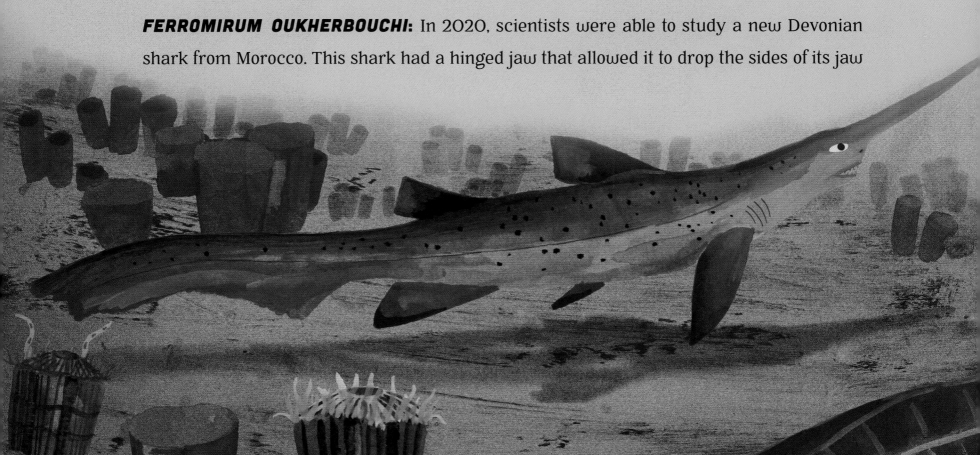

downward to reveal larger, sharper teeth hidden inside the mouth. Scientists think this might have been a precursor to the rotating teeth of modern-day sharks.

FRILLED SHARK: Frilled sharks are called living fossils, because they haven't changed much in 80 million years. These strange deep sea sharks can be up to 7 feet long and have 300 teeth!

PTYCHODUS: Ptychodus is a mysterious shark from the late Cretaceous. Scientists think they might have been over 30 feet long, and they had thick flattened teeth that could crush mollusks and other hard-shelled animals.

SAIVODUS STRIATUS: Also in 2020, scientists in the Mammoth Cave system in Kentucky found a giant shark jaw fossilized in the rock. This top predator, which was probably bigger than a great white, lived in the Carboniferous era. They also found fossils of smaller eels and sharks that might have fed on the Saivodus carcass!

SARCOPRION: This cousin of Helicoprion lived during the Permian period in Greenland. It had a small, compact tooth whirl and a long, pointed snout, and scientists think it might have hunted faster squid and fish prey, the way today's mako sharks do.

SIXGILL SHARK: The ancient-looking sixgill first showed up in the early Jurassic period, and it's still around today. In fact, sixgills and their relatives are the oldest known family of modern sharks!

XENACANTHUS: This small freshwater shark was a relative of Orthocanthus and survived until the end of the Triassic period. At only 3 feet long, it could have hunted in waters too tangled and shallow for its bigger cousin.

WAYS TO HELP SHARKS AT HOME

- Get Educated! Check out sites like sharks4kids.com to find out more about sharks and how to help them.

- Encourage your family to eat sustainable seafood—which is caught or farmed responsibly.

- Reduce plastic trash in your home when you can.

- Pick up trash when you see it, especially at the beach.

- Support shark conservation organizations like Shark Advocate International.

- Finally, tell people the good things you learn about sharks. Many people think sharks are mindless eating machines like in the movies, but sharks are important to the ocean. You can help sharks by learning more about them and sharing the things you learn!

RESOURCES

Bonner, Hannah. *When Fish Got Feet, Sharks Got Teeth, and Bugs Began to Swarm: A Cartoon Prehistory of Life Long Before Dinosaurs*. Washington, D.C.: National Geographic Society, 2009.

Ebert, David A., and Sarah Fowler. *Sharks of the World*. Princeton, NJ: Princeton Universtiy Press, 2021.

Eilperin, Juliet. *Demon Fish: Travels Through the Hidden World of Sharks*. New York, NY: Pantheon Books, 2012.

Ewing, Susan. *Resurrecting the Shark: A Scientific Obsession and the Mavericks Who Solved the Mystery of a 270-Million-Year-Old Fossil*. New York, NY: Pegasus Books, 2017.

Howard, Abby. *Ocean Renegades!* New York, NY: Amulet Books, 2018.

Klimley, A. Peter. *The Secret Life of Sharks*. New York, NY: Simon & Schuster, 2003.

Shiffman, David. *Why Sharks Matter*. Baltimore, MD: John Hopkins Press, 2022.

Skomal, Greg. *The Shark Handbook, 3rd Edition*. Kennebunkport, ME: Cider Mill Press, 2021.

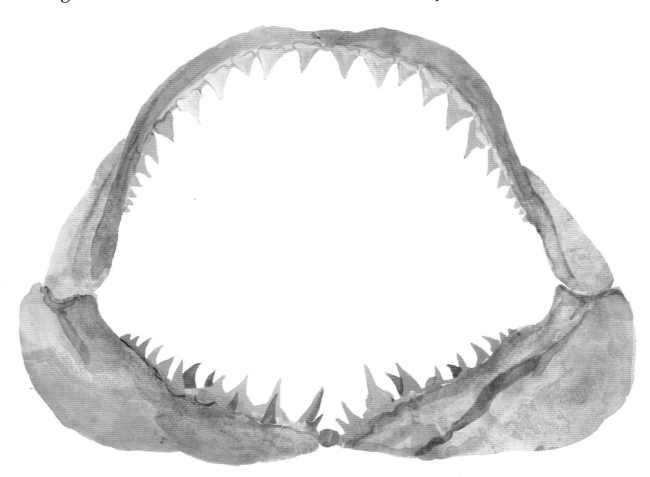

For Alex, always —M.F.

For my Dad, Jim, and his love of the sea —G.W.

Artwork painted by hand in gouache, scanned, and then put together digitally.

Cataloging-in-Publication Data has been applied for and
may be obtained from the Library of Congress.

ISBN 978-1-4197-4773-1

Text copyright © 2022 Miriam Forster
Illustrations copyright © 2022 Gordy Wright
Book design by Heather Kelly

Printed and bound in China
10 9 8 7 6 5 4 3 2 1

Abrams Books for Young Readers are available at special discounts when purchased
in quantity for premiums and promotions as well as fundraising or educational use.
Special editions can also be created to specification. For details,
contact specialsales@abramsbooks.com or the address below.

Abrams® is a registered trademark of Harry N. Abrams, Inc.

ABRAMS The Art of Books
195 Broadway, New York, NY 10007
abramsbooks.com